HOW TO SPEAK CHICKEN

Melissa Caughey

Storey Publishing

The mission of Storey Publishing is to serve our customers by publishing practical information that encourages personal independence in harmony with the environment.

Edited by Deborah Burns and Lisa Hiley
Art direction and book design by Alethea Morrison
Cover photography by © Jutta Klee/Getty Images, front; © Global IP/iStockphoto.com, front inset; © Image Source/Getty Images, back; © Jared Leeds, author
Interior photography by © Jared Leeds, 6, 9, 24, 61, 105, 106, 117, 123; © Mars Vilaubi, 27, 37 top, 81, 82, 86, 114, 134, 139; and © Melissa Caughey, 43, 57, 67, 129, 140
Additional interior credits appear on page 144
Illustrated patterns by © Yao Cheng Design

Storey Publishing
210 MASS MoCA Way
North Adams, MA 01247
storey.com

Printed in China by R.R. Donnelley
15 14 13 12 11 10 9 8 7

Library of Congress Cataloging-in-Publication Data

Names: Caughey, Melissa, author.
Title: How to speak chicken : why your chickens do what they do & say what they say / by Melissa Caughey.
Description: North Adams, MA : Storey Publishing, 2017. | Includes bibliographical references and index.
Identifiers: LCCN 2017030616 (print) | LCCN 2017041112 (ebook) | ISBN 9781612129129 (ebook) | ISBN 9781612129112 (pbk. : alk. paper)
Subjects: LCSH: Chickens–Behavior. | Chickens–Psychology.
Classification: LCC SF493.5 (ebook) | LCC SF493.5 .C38 2017 (print) | DDC 636.5–dc23
LC record available at https://lccn.loc.gov/2017030616

This book is dedicated to my husband, Peter, who is continually encouraging me to spread my wings.

Preface

The things you are passionate about are not random, they are your calling.

Fabienne Frederickson,
author of *Embrace Your Magnificence*

What started out as a journey to teach my children about sustainability and responsibility, while providing our family with fresh eggs from our own backyard, has turned into something magical. If you are a backyard chicken keeper like me, you absolutely love your flock. We spend countless hours outside with our girls and can't wait to get home from work and school to be with them. Over the years, I have tried to see my flock through both the heart of a mother and the magnifying glass of a scientist. In that time, I have developed such close bonds with my chickens that I believe I have been accepted into their fascinating world.

Long before I dreamed of owning chickens, I trained as a nurse practitioner, studying sociology, anthropology, psychology, and biology. Learning about interpersonal relationships, culture, language, and societal roles in the melting pot of Los Angeles opened my eyes to the world. At the age of 20, I had no idea that my skills in understanding cultures and my respect for those who are different from me would come in handy in a chicken coop on Cape Cod.

Scientists use many tools and methods to research flock dynamics, including vocalization playbacks, video recordings, 3D animation, wireless backpack microphones worn by chickens as they interact, and even robotic chickens! I don't have any of these sophisticated research tools, but with an open mind and an open heart, I have come to many of the same conclusions as the scientists simply by spending time with my flock and paying close attention to them. And I might have an advantage over the scientists in knowing my hens so well — I believe they have made me an honorary flock member.

This book is a culmination of my experiences in keeping chickens, which have allowed me to gain insight into their communication, body language, intelligence, social interactions, emotions, and problem-solving abilities. I am here to tell you that they are clearly not "birdbrains," and I invite you on a journey to explore how their minds, bodies, and emotions work. Their world is pretty amazing!

I encourage you to take the time to watch and listen to your chickens chattering as things are happening. Repeat what you hear back to them and watch their response. At first some will stop in their tracks, and like dogs, cock their heads from side to side as they stare at you with their intense chicken eyes. They're surprised that you can speak chicken! Keep on copying them. Repeat the sounds they make as you open the coop, listen to what they say when you throw out scratch, follow them around the yard as they explore, and even follow them into the coop to say good night. Repeat the coos and clucks and buk-gaws, and before you know it, you too will be fluent in chicken speak. Who knows, you might even find out that you have a chicken name!

1

Chicken Translator at Your Service

Understanding What Your Chickens Are Saying

Humans often communicate without words: we use our eyes, gestures, facial expressions, and body language. Chickens use most of these, but they have also more than two dozen different vocalizations. These calls include ones relating to territory, mating, distress, danger, fear, happiness, discovery of food, and nesting.

Contrary to what one may hear from the industry, chickens are not mindless, simple automata but are complex behaviorally, do quite well in learning, show a rich social organization, and have a diverse repertoire of calls. Anyone who has kept barnyard chickens also recognizes their significant differences in personality.

Dr. Bernard Rollin, author of *Farm Animal Welfare*

It is said that if you want to really learn a language, you should move to a place where that language is spoken. To learn to communicate better with my chickens, I simply took a small stool out to the flock and perched with them, day after day, quietly paying attention and noting their interactions. Over the course of their lives, I take the time to get to know each one as an individual. Some seem to enjoy hanging out with people, while others keep to themselves unless treats are being offered, but they all "talk." I learned by emulating their coos, clucks, and squawks, paying careful attention to the intonations. It must be a bit like learning Mandarin Chinese, where different inflections give the same word or sound different meanings!

Animals speak volumes if you take the time to listen.

LEARNING CHICKEN LANGUAGE

I'VE ALWAYS TALKED TO MY FLOCK while doing my chicken chores. One day I realized that, interspersed with English, I was mimicking the girls. That got me thinking: Could I really speak chicken? Some folks thought I was crazy, but as I started listening more consciously, I discovered that I knew plenty of their calls — danger signals, happy calls, hellos, good nights, and even encouragement chatter. I was learning to speak chicken!

Over the next few months the entire family practiced with the flock, copying and trying to interpret their calls. Then one day my kids and I were visiting another flock. As we sat and watched, the flock ignored us and went about its usual chicken business. I decided to try something fun. Truthfully, I was a little embarrassed to share the fact that I was studying the language of the flock. I think my friend thought I was crazy, but she indulged me.

When we used chicken language to greet them, the response was astonishing. Their heads popped right up, as if to say, "Hey, is there another chicken here?" Some came over to us out of curiosity, and the rest went back to chickening. We "chatted" a bit more with them using other intonations and sounds. Then for the final test, I used the alarm call to alert them to danger coming from above. Every single chicken paused and became statuesque with one eye to the sky. Just as I thought, they understood me! That was the moment I learned that I indeed was onto something.

IN THE BEGINNING: BABY TALK

Mother hens communicate with their chicks via clucks and squawks, even when they are still nestled in their eggs. It starts with the chicks. Twenty-four hours prior to hatching, a peeping sound, also known as "clicking," is heard from within the egg. This sound serves as communication to the mother hen from the babies, as well as among the chicks. As the hen answers back, the peeps inform her how long to stay on the nest and how many babies to expect.

Once the chicks have hatched, the mother watches over her babies, teaching them where to find food and what things are good to eat. She alerts them to danger and guides them to safety. She gathers her brood into the warmth of her downy feathers with a special call.

Hens are very caring and nurturing to their young. In one study, researchers used puffs of air to "irritate" newly hatched chicks. You can imagine the mother hen's response. She was not happy! Even though she herself did not mind being puffed with the air, when her chicks were upset, she became upset as well.

While the chicks are still nestled in their eggs, the broody mother talks to them. I can only imagine she is bonding with them via her quiet coos, clucks, and mutterings. She is their first cheerleader in life.

BROODER VOCABULARY PRIMER

Until about 14 weeks of age, chicks have a specific language consisting of peeps, cheeps, and pleasure trills of varying tones and volume. These sounds are used only by chicks. I believe that hens understand their chicks, but they do not use or mimic sounds made by the chicks. This early language can represent feelings of happiness, danger, or comfort. Whether the chicks are hatched by their mother or in an incubator, this chick language is universal.

Chicks primarily make this contented purring noise as they fall asleep. (Think of a happy cat.)

Sweet Peeps These are the soft sounds of casual conversation as the chicks busily explore their surroundings and learn about the world.

Discovery Peeps Excited, repetitive chirps mean "Look what I found!" At the same time, the chick is probably trying to keep her siblings from stealing her find, which is usually a tasty morsel.

Distress Peeps Loud, strong, continuous peeps can indicate pain, stress, or an alert to the others that something is wrong. You'll hear them when chicks are scooped up from the brooder, separated from their flockmates, can't find their mother, or are feeling cold.

All about the Crowing

When your chicks are about eight weeks old, you might be surprised by a very pathetic sound coming from the brooder early one morning — like someone clearing his throat while trying to shout. This is the sound of a young rooster (cockerel) trying to crow for the first time. If you are lucky enough to see him, he puts every ounce into this effort, digging deep to make the sound while standing tall, puffing out his chest, and channeling the crow from the tips of his tail feathers to his vocal chords.

If there is more than one cockerel in the flock, it is common for those below the head cockerel not to crow or openly mate with the young hens (pullets). When a group of roosters are raised together, the dominant male is determined early. They coexist peacefully, with the subordinate roos "flying under the radar" (sometimes even from the keeper!). The other roosters will usually only assert themselves and show their male characteristics once the dominant rooster leaves the flock.

As I always tell fellow chicken keepers, you cannot fully determine if a chicken is a rooster or hen until he crows or she lays an egg. Some breeds are trickier than others.

GROWING UP: LOOKING AT BODY LANGUAGE

In addition to having a vocabulary (more about that on page 34), chickens have nonverbal methods of communication.

Pecking

The beak is the best tool that chickens have. Chickens use their beaks for the same reason we use our hands: for communication, protection, exploring, eating, grooming, and so many other day-to-day activities.

Peck first, ask questions later.

Squatting

This behavior is specific to hens. They squat down and stretch out their wings. It is a submissive behavior often displayed to roosters to accept their willingness to mate, but it may also be seen when a person reaches for or picks up a hen.

Tidbitting

Tidbitting is a dance that a rooster performs to attract a mate. He bobs his head up and down while picking up tiny bits of food, dropping them, and repeating. During this time the rooster's bright red comb (flesh on top of his head) and wattles (flesh under his chin) move about.

Roosters are romantics at heart. It is not uncommon to find them fluffing or sitting in the nests, sitting on eggs, or partaking in chick rearing.

Researchers have discovered that hens prefer male roosters with larger combs. They believe it is the rooster's comb that makes him an attractive mate to the hens during tidbitting.

A tidbitting rooster goes all out to impress his chosen hen.

TIPS FOR WATCHING YOUR FLOCK

Wear plain clothing. A word of caution — chickens love to peck at bling and dangling objects. (They'll investigate freshly manicured toes, too!)

Be quiet and stay still. Your voice and movements will distract them from behaving naturally.

Be patient. At first the chickens focus on you, but given time they'll go about their business.

Spend plenty of time. It takes many hours of observation for the flock to treat you as an honorary member.

Mornings are a good time to watch the flock as they go about their daily rituals of eating, drinking, and exploring. When it comes to taming or training individual birds, wait until the afternoon, when they are less distracted.

Watch how they behave with and without treats present.

Observe your birds in a location where they feel safe. Do not try to observe them when they are in a new place or situation. Keep potential perceived predators away, such as dogs or strangers.

Be aware of the weather. You may notice different behaviors in different conditions and seasons.

Keep a journal with your observations and photos or sketches. Who knows, you might end up discovering something new!

Record what they are saying with your smartphone so you can play it back to refresh your memory later. This will help you interpret chicken speak!

You will learn more observing your flock if you are quiet.

COMMON BEHAVIORS TO LOOK FOR

Dust Bathing

Chickens keep clean by wallowing in dirt and dust and working the particles down to the skin. This process helps to control pests and parasites such as mites and poultry lice. Scratching away, sometimes working collaboratively in a group, a chicken digs a hole large enough to sit in, sometimes even a bit bigger so a friend can join in. Once the dust bath is prepared, the chickens roll, stretch, and toss the dirt into the air so it falls onto their bodies. I think of them as "boneless chickens" as they contort into yoga poses.

Sunbathing

Chickens love to bathe in the sun. It is not uncommon to see them lying on the ground or in their dust-bathing holes with their wings unfurled so their feathers can soak up the sun. Sometimes they relax so much that they drift off to sleep.

Preening

After dust bathing, the chicken coats her feathers with oil to help repel water, keep the feathers clean, and prolong the life of each feather. With her beak, she collects a bit of oil from the uropygial gland that sits above her tail and smooths the oil along her beautiful fluff.

Beak Rubbing

Chickens rub their beaks on hard objects such as logs in order to sharpen, shape, and clean them. Left unmaintained, beaks grow just like your fingernails.

Egg Laying

A hen can lay an egg approximately once every 26 hours. This ability starts at around 18 to 20 weeks of age and may continue for the entire life span of the hen. The number of eggs begins to taper off at about 2 years of age. Hens require a safe place to lay their eggs, proper nutrition, and 14 hours of daylight to stimulate egg production.

Roosting

Chickens go to bed at dusk. They have a natural tendency to roost off the ground, and in the absence of a coop they will take to the trees. In cold weather, you will find them snuggling together on the roosts to keep warm.

THE BROODY HEN

All hens lay eggs, but "broodiness," or the desire to hatch eggs, is a trait that has been bred out of most breeds for the simple fact that when hens are brooding, they do not lay eggs, which isn't good for profits. But some hens still have a particularly strong desire to be mothers. Silkie Bantams come to mind, as many of them are perpetually broody. I love Silkies for this reason, but I don't recommend them to folks who want to keep chickens solely for their eggs.

A broody hen leaves the nest just once a day or so for three weeks.

A broody hen will lay eggs every day but won't walk away from them like most hens do. If you are lucky enough to have a broody hen, you will quickly discover that she will happily sit on her own eggs, another hen's eggs, wooden eggs, or even no eggs. She will also go to great lengths to hide her eggs from any threat, including the chicken keeper who comes to collect them. Some hens manage to disappear for weeks, only to return to the chicken run with a line of baby chicks behind them.

Chicken eggs take approximately 21 days to hatch, and the broody hen incubates them under her downy fluff pretty much around the clock. Once she lays the "right" number for a clutch,

A clutch is rarely an even number because an odd number of eggs fits together better.

which is typically 9, 11, or 13, she begins plucking her chest feathers to create a patch of exposed skin, called the "brood patch." She uses the feathers to line the nest, then presses her brood patch directly on the eggs. The skin contact provides the eggs with the proper temperature and humidity for chick development.

The hen sits close to her eggs, flattening her body and wings for the most coverage of the nest. She rotates the eggs with her beak in the nest to ensure proper positioning and development as well as access to the brood patch. A protective mother, she coos to the eggs but will growl at fellow flock members and even chicken keepers to warn them to stay away.

Chicks know to snuggle into their mama's feathers to keep warm.

Once every 24 hours, she leaves the nest for no more than 20 minutes to eat, drink, and pass one huge, smelly broody poop. (Chickens usually poop a lot — up to 50 times a day! They even poop at night when they are sleeping.) Otherwise, day in and day out, she sits in a trance on the eggs, waiting for her babies to arrive.

Silkie Bantams have a particularly strong desire to be mothers and many are perpetually broody.

MEETING DOLLY'S BABIES

DOLLY, ONE OF OUR SILKIE BANTAM HENS, was so perpetually broody that we called her the Dolly Mama. At the time we had a rooster. We knew we needed to rehome him, but before we did, we let Dolly hatch out a clutch of fertilized eggs. For three weeks, she diligently perched on her eggs, flattened out like a pancake and growling at anyone who came near her.

A couple of days before they were due to hatch, I heard peeping from the eggs! Dolly still sat on the nest, cooing to her babies and coaxing them to start pecking. I checked her frequently for the next couple of days, anxiously waiting to welcome her chicks into the world.

Finally, I saw one little baby tucked under her wing. When I gently lifted her up, I could see faint hairline cracks circling the middles of a few other eggs, carefully pecked away by the egg tooth on each baby chick. By evening, we could see three more babies under her wings, snuggled in her chest feathers. By the end of day two, eight babies had successfully hatched.

Dolly was an amazing mother. I admired her dedication as she taught her chicks all about life over the next six weeks. Once her brood was fully feathered and knew all they needed to know to survive among the flock and in the great big world, she set them on their way. Within two months, she was broody again, ready to hatch some more eggs!

CHICKEN LANGUAGE EXPLAINED

Chickens use their language to communicate about food, to warn of predators or other threats, and to converse among themselves about what's going on. This "grown-up" language, used by all chickens, starts at about 14 weeks of age.

The *buh-dup* greeting sound is heard among members of the flock as they come and go during the day. They also say it to humans.

When feeling threatened on her nest, a broody hen makes a sort of a yell that trails into a grumble. It sounds like a tiny dinosaur roaring!

Buh-dup

Crowing

It is a fallacy that a rooster crows only in the morning. He will crow 24/7 to announce his presence and warn of perceived threats.

Alert

Grrrrrr, buk, buk, buk, buk, buk, buk BUKGAW

Chickens make an alert call to signal that danger is near. It starts out as a low, rolling growl and progresses in volume and intensity into a full-out yell. There are two alert calls. The low-pitched alert for a ground predator is accompanied by the chickens running for cover. The higher-pitched alert for a threat from the sky causes the chickens to pause, crouch, and look upward.

Dispute

(General squawking and squabbling)

A burst of noisy carrying on usually means a disagreement is occurring. The flock is not getting along. Typical squabbles may involve a tasty morsel of food, a favorite roosting place, or sorting out a perceived spot in the pecking order.

Bwah, bwah, bwah, bwah

This call sounds a bit frantic and slightly impatient. It is usually made by a hen while pacing outside her favorite nesting box when it is occupied by another hen. Once inside the coveted nesting box, she quiets down and will not "speak" until she lays her egg.

Buh-gaw-gawk, buh-gaw-gawk, buh-gaw-gawk

Egg Song

Buh-gaw-gawk, buh-gaw-gawk, buh-gaw-gawk

This distinctive egg song announces that a hen has laid her egg for the day. It's the only time a hen will be really loud, even louder than the alert call. She starts the call as she leaves the nesting box and continues to celebrate her accomplishment as she returns to the flock. It's not uncommon for other hens to join in. They seem to enjoy celebrating one another's successes.

The soft, airy-sounding good-night murmur is like an evening roll call that means, "Yes, I'm here and I am okay."

HOW CHICKENS SAY GOOD NIGHT

ONE WINTER EVENING DURING A BLUSTERY SNOWSTORM, I reluctantly made the trek to the coop to lock up the chickens, knowing it had to be done to keep them safe from lurking predators. I entered the coop and stood for a moment to catch my breath from the wind and chill. As I felt the snow melting on my face, I could hear the chickens calling to one another from the darkness of the roosts with these sweet *doh doh doh* sounds.

I stayed still, taking it all in and trying to figure out just what was going on. One by one they called out to each other, as if taking a head count. I listened to the calls for quite a while. Then I chimed in, mimicking what I was witnessing. To my surprise, they answered back. I realized this is how chickens say good night. It is their nighttime ritual.

CHICKEN TALK

Chickens have been part of our lives for so long that it's not surprising that many of their behaviors have found their way into our language. Perhaps we aren't so different after all!

Pecking order

Henpecked

Fussing like a mother hen

Crowing about an achievement

Putting all your eggs in one basket

Chickens coming home to roost

Counting your chickens before they hatch

Mad as a wet hen

Building a nest egg

BOTH NURSING HOME AND NURSERY

A **COUPLE OF YEARS AGO,** we introduced some new chicks to our existing flock. Our older girls were in "henopause" and rarely laying eggs. As the chicks, mostly Easter Eggers, came of egg-laying age, they began to spend time in the nesting boxes, rearranging the pine shavings with their beaks and claws, and sitting fluffed out in the boxes as though practicing for the big day when they would lay their first eggs.

As the new girls sat in the boxes, I was surprised to see the older girls visiting them. They'd crane their necks into the boxes and peer inside. The girls would converse, sharing words of encouragement. And so it went until finally one of the new girls,

Egg laying is always cause for celebration.

Cuddles, laid her egg. Every single hen came in to see the egg and repeated Cuddles's egg song, as though celebrating her accomplishment.

Within a few weeks, all the new girls were laying eggs. One day, as I went to harvest the eggs, I discovered one of the older hens, Oyster Cracker, in the nesting box perched on a freshly laid clutch. It was a magical moment seeing her in that nesting box with those eggs. She hadn't laid an egg in years, yet her mothering instinct kicked in again and she was ready to add to the flock, too. I was completely surprised when Oyster Cracker did in fact lay an egg. Afterward, she peered down at it between her legs for what seemed an eternity, apparently as shocked as I was!

Oyster Cracker and her surprising egg – the black spot in the lower corner is Cuddles looking in on her friend!

DO YOU HAVE A CHICKEN NAME?

As I studied my flock, the meaning of one particular sound always eluded me. I had been hearing it for years, starting about six months into our chicken-keeping adventure. I could not pin down what it meant, despite watching our birds for so long. I simply dismissed it as another chicken greeting, but I learned that I was wrong. It never occurred to me that they might have been watching me, too, until I had the pleasure of meeting an incredible woman named Sy Montgomery.

Many bird species communicate with a variety of sounds. Intelligent birds like parrots and chickens even use specific sounds to identify individuals.

Sy has spent her life traveling the far reaches of the world seeking out the secrets of the animal kingdom, but as she is also a devoted chicken keeper, we bonded instantly. She told me that it is not uncommon for animals to have distinct calls for different individuals — names, if you will. The use of specific names has been discovered in various mammalian species, including monkeys, whales, and dolphins, but has been especially studied in birds, particularly parrots. Could this puzzling vocalization be a name that the chickens had given to me?

I had been interpreting this call as another greeting, but when I paid closer attention, I realized it was in fact my chicken name! Tilly, my original head hen, had named me. She was always the most vocal. The first to greet me every morning, she would often find me in the garden, seeking me out for treats or just wanting to spend time with me, her chicken mom. After reflecting on this epiphany, I was in awe. I had a chicken name.

To me, my chicken name sounds regal. The first three sounds are low, and the last one is almost an octave higher. *Bup, bup, bup, baaahhhh.* They use this sound when they see me, when they want something from me, and when they are snuggling in my lap.

Tilly, Oyster Cracker, and many other members of my original flock have passed on, but my new girls, especially Fluffy, still use my chicken name and carry on the tradition. It makes me smile every time I hear it. I think of my name as a gift from Tilly, like the gift we give to our own children when we choose a name for them.

FLOCK TALK

SY MONTGOMERY

Award-winning author of *The Good Good Pig, Soul of an Octopus,* and *Birdology*

Rural New Hampshire

Sy Montgomery has traversed the globe to study the connections we share with our fellow creatures. She has been chased by an angry silverback gorilla, bitten by a vampire bat, and hunted by a tiger. She has also has swum with piranhas, electric eels, and dolphins.

Current flock

Three black Australorps and six Dominiques

Favorite chicken

Omeletta, who was was renamed when she underwent a hysterectomy to clear up problems related to a diseased shell gland. Hers may have been the first hysterectomy ever performed on a hen in the history of veterinary medicine.

Best thing about chickens

Chickens are fabulous ambassadors for their kind. Our hens are so sweet, so appealing, so obviously curious and clearly intelligent that everyone who meets them is charmed. All I have to do is introduce them. The hens do the rest.

I also love flock etiquette and flock dynamics. Our flock has always been a peaceful feminist utopia. But at one point we had a tenant who kept another flock on our property — and they were ready to rumble! Even though the other flock was fenced in, our free-ranging ladies got so upset they mostly moved next door. They just didn't even want to see those other hens!

And I love when they catch sight of me when we've been apart. They rush to greet me like I am an international rock star! There's no better boost to an ailing ego.

How to Behave in the Henhouse

Rules, Etiquette, and Social Graces

Like any other group of animals, a flock of chickens has rules that ensure a smooth existence. Every hen knows her place and her duties, and every rooster his, and for the most part all respect the internal hierarchy, or pecking order. This instinctual behavior on the part of chickens offers safety in numbers, increases the chances of survival, and ensures the continuation of their genes, which is the ultimate goal of all animals.

Chickens are capable of reproducing at around 20 weeks of age, and they can have a large number of babies in a short period of time. The precocious chicks are independent about six weeks after hatching. Chickens are omnivorous scavengers, and they quickly learn how to scratch through leaf litter, eat the best plants, and hunt insects, slugs, worms, and even small reptiles and mammals, such as mice. By the time they are independent, they have also learned the lessons of flock behavior.

Scientists have learned that this bird can be deceptive and cunning, that it possesses communication skills on par with those of some primates and that it uses sophisticated signals to convey its intentions. When making decisions, the chicken takes into account its own prior experience and knowledge surrounding the situation. It can solve complex problems and empathizes with individuals that are in danger.

Carolynn L. Smith and Sarah L. Zielinski,
"The Startling Intelligence of the Common Chicken"

There are an estimated 50 billion chickens to the world's 7 billion humans.

A BRIEF LESSON IN "HENTHROPOLOGY"

The history of chickens is a mystery that is still unraveling. Scientists believe, on the basis of DNA studies, that the modern domestic chicken (*Gallus gallus domesticus*) originated when red jungle fowl (*G. gallus*) crossed naturally with gray jungle fowl (*G. sonneratii*) about 8,000 years ago. Research indicates that this happened in a few areas of Southeast Asia, including southern China, India, Burma, and Thailand.

Archaeological digs in China show the earliest suggestions of the presence of domesticated chickens within the cultures of Cishan (5400 BCE), Beixin (5000 BCE), and what is now Xi'an (4300 BCE). It was through cockfighting that chickens began to spread around the globe. Skeletal remains and other evidence of

Gallus gallus

Gallus sonneratii

chickens have been found in Europe, Africa, and the Polynesian Islands. Some researchers even think that chickens arrived in Chile from Polynesia, not via the Spanish conquistadores, as previously believed.

Though chickens were long kept as a source of eggs and meat, it wasn't until the late nineteenth century that people began developing breeds with specific traits, such as size and egg-laying ability. In 1874, the American Poultry Association introduced the *Standard of Excellence* (now published as the *American Standard of Perfection*) for chicken breeds. Many years of selecting for certain characteristics have resulted in at least 350 recognized breeds around the world.

Standards of breed perfection from the late nineteenth century

MANNERS AND THE PECKING ORDER

The pecking order is the social hierarchy that exists within the flock. The concept was first described in 1921 by Thorleif Schjelderup-Ebbe, a prominent Norwegian zoologist and comparative psychologist who as a boy was fascinated with the chickens on the farm where he spent his summers. Over a number of years, Thorleif spent hours immersed in the flock, studying,

Pecking order, more broadly referred to as dominance hierarchy, is a factor in many animal societies.

watching, and learning. He kept detailed notebooks filled with his observations. Years later, Thorleif returned to those records when he wrote his PhD dissertation.

Thorleif originally described the pecking order as an "expression of dominance," which he measured by two things: a chicken's ability to ward off another chicken's aggression and the level of aggression accomplished by the beak. He noted that the head hen will often peck any subordinate hen that she feels is infringing on her preferred living space or nesting box, or her access to food. The subordinate hen will then peck a lower one and so forth until the bottom of the order is reached.

Having first access to food is one of the benefits of being at the top of the pecking order.

In a flock of hens, the pecking order is mostly sorted out over food. As hens move to the top of the pecking order, there are often displays of fluffing, squawking, and strutting. Those lower in the rank will often refuse to fight with the higher-up hen. They simply submit to their subordinate position.

The purpose of the pecking order is to prevent aggressive fighting within the flock. Avoiding such disputes cuts down on the amount of energy and resources used for competition. A wild flock with a well-established hierarchy can focus on finding food, mating, and raising broods. In a domestic flock with a single rooster, only the hens are in the pecking order. The rooster serves as the leader and protector and is outside the order, though if there are several roosters in a large flock, they will establish their own hierarchy.

WHAT IT TAKES TO GET TO THE TOP

Pecking order position is not always based on size. Some believe that the pecking order is based on the breed, personality, and intelligence of the hens. Often hens just submit to their position after observing those above them, choosing this option instead of squabbling. It's remarkable that some chickens, through their own observation, can size up the abilities of the other flock members, recognize their own strengths and weaknesses, and determine whether or not they have the ability to take on a chicken higher up in the pecking order.

It seems to me that the pecking order might also depend on trust, respect, and possibly fear, concepts that may seem foreign to our understanding of bird interaction. But it is not in a flock's interest to have internal strife. Every individual must understand who is in charge and must agree to get along.

Flock Size Matters

In the wild, flocks typically do not go above 20 members, and experience with domestic birds shows that a pecking order works best in flocks of fewer than 20 chickens. Within a flock of this size, the chickens are able to recognize and keep track of one another. Larger flocks exhibit much less stability, more aggression, and far more disputes. Sometimes a few chickens will form a subgroup within the larger flock, but they are still subject to the rules of the bigger group.

A flock that has plenty of space and food is more likely to establish a stable pecking order.

MY MINI FLOCK

ONE OF MY FAVORITE BREEDS is the Silkie Bantam. I guess I first fell in love with them because of their appearance. They were so different with all that fluff, and they were so small, too. With 5 Silkies in my flock of 10, I soon realized that they also had super-sweet personalities, were great mothers, and did not care to be high up in the pecking order. They all seemed happy to just hang out together and ignore the others; in fact, it appeared that they formed a mini flock within the larger flock.

It is not uncommon to see smaller flocks within larger flocks. This can sometimes occur with birds of the same breed in a mixed flock, as it did in this case. My Silkies stuck together. They ate together, explored together, laid eggs together, took dust baths together, and roosted side by side away from the rest of the flock.

They also seemed to squabble just among themselves. They established their own pecking order that was completely unrelated to the larger flock. Where each Silkie fell within this mini flock seemed to carry more weight than her position in the larger flock. Everyone coexisted, two flocks in one.

Dolly, our sweet broody Silkie

PECKING ORDER IN THE BROODER

Sorting out the pecking order starts just days after the chicks hatch. Once they feel comfortable in the brooder, their temporary home during the first few weeks, and know where to find the food, the water, and the best sleeping places, the chicks begin to figure out who they are. At first, there can be a bit of chasing, but within a few days, a dominant female chick will begin to assert herself.

She is braver and more curious than the others; she is unafraid to taste new items like mealworms or clumps of fresh grass. She has first dibs at the food and keeps others away from the food dish until she has had her fill. She also sleeps in the best places. She is almost always smarter than the others, and possibly but not necessarily the biggest in the flock. This chick will most likely end up as head hen.

If there is more than one male in the brooder, you might notice some full-on showdowns during this time. Two male chicks will face off and walk in a circular fashion while staring each other down. They stretch out their necks and bump chests until one comes out the winner and the other accepts a lower position.

Unlike the process of introducing adult chickens to an existing flock, figuring out the flock hierarchy in the brooder is a relatively peaceful process, especially among female chicks. Things among the girls are quietly understood. Once established, this initial pecking order is respected through the life of that specific flock. Even when the head hen becomes old and unable to perform her duties, she will receive the respect she earned all those years ago when the chicks were merely days old.

However, if the makeup of the flock changes, due to changes in health, the loss of members, or the integration of new chickens into the flock, the pecking order will adjust.

The dominant female chick will begin to assert herself shortly after hatching.

POSITIONS WITHIN THE FLOCK

Within the flock there are a few key functions. In a large flock, the chickens that do not have specific roles sort themselves into their own ranking. When a hen can no longer perform in her specific role, she is quickly replaced by another. Perhaps the replacements were their understudies all along.

- As chief protector, he watches out for danger.
- He is above the hens' pecking order and has the ultimate say.
- He keeps the hens in line and intervenes in squabbles.
- He has his favorite girls, independent of where they fall in the pecking order.

Head Hen

- She is most often the healthiest, strongest, and smartest hen.
- She leads the flock and controls where they free-range.
- She is usually the first to access food, though she may sometimes allow others to eat before she does.
- She keeps order, with or without the presence of a rooster.

- The other hens often learn from her and follow her lead when exploring, operating new equipment such as feeders and waterers, learning about new foods, and so on.
- She is often the last one to enter the coop at night, making sure her entire flock is home before she roosts.

Sentinel

- In the absence of a rooster, the role of sentinel is assumed by a hen other than the head hen. There is usually only one official sentinel, although the rest of the flock tends to be on alert much of the time.
- The sentinel keeps guard with one eye on the ground and one to the sky, monitoring for predators.
- The sentinel often finds an elevated spot that serves as a lookout point.
- When a threat appears, the sentinel alerts the flock. That alarm call, notifying them whether the attack is from the sky or the ground, is then carried through the flock by the others.

Bottom of the Order

- The bottom of the order is sometimes occupied by a single chicken of a different breed or one with a submissive personality.
- Some birds seem to care more about pecking order ranking than others. A breed such as the Polish tends not to be dominant; a breed such as the Silkie is indifferent to the pecking order.
- The bird at the bottom of the order might be either weak or impaired in some way.

BULLYING AND OTHER BAD BEHAVIOR

Chickens do display bad behaviors, as do most other animals and certainly some people. Some chickens just are not nice. Whether it is a hen or a rooster, a bully can turn on other flock members or even their caretakers.

Chickens can become aggressive for a number of reasons. When I talk about aggression, I don't mean a squabble over a worm, a favorite nesting box, or roosting space. I am talking about malicious and continuous pecking that can draw blood, cause injuries, or even lead to death. This may exist outside the pecking order (see There's One in Every Bunch, page 66).

Some chickens are aggressive by nature, but most often this behavior arises from overcrowding, nutritional deficiencies, illness, and other poor living conditions. Ideally, every confined chicken should have at least 10 square feet of space outside the coop and 2 to 4 feet of floor space inside the coop. Chickens should also receive a high-quality feed, with fresh food and other treats limited to 10 percent of their diet. Chickens benefit greatly from being allowed to forage free from their coops and runs on a regular basis. Flocks that free-range all day, every day, have fewer bullying issues.

Factory-raised hens are kept in close quarters from the time they are chicks. They often have their beaks removed to prevent pecking issues, although many wind up almost completely featherless anyway. If they had more space, excessive pecking would go away. This is among the many reasons activists are advocating for more floor space in farms per chicken. Some are lucky enough to be rescued. It is a sight to see these poor hens exposed to open pasture for the first time: they revel in scratching the dirt, looking for bugs, taking their first dust bath, and finally being be free.

Feather pulling is often seen in stressed, overcrowded flocks.

MEAN ROOSTERS

I think roosters get a bad rap. A number of people who have raised a rooster from a day-old chick have told me that the rooster was their favorite early on. But many roosters change at around six months of age. They become aggressive, chasing their owners and even sparring at them from behind.

Little do folks realize that the rooster is only doing what comes natural to him. He is possessive of his girls. When his hens go into the submissive squat for mating and you or your children pet them or pick them up, he sees this as you trying to mate with his girls. He is just trying to prevent that from happening.

This is often the point when roosters from backyard flocks are rehomed or added to the soup pot. The risk of injury to people becomes too great. Some flock owners insist that they can change a rooster's behavior by asserting their own dominance. Techniques might include holding the rooster on his back, diverting bad behavior by shaking a noisemaker such as a can full of rocks, or physically fending off attacks with a rake or broom.

Unfortunately in our situation, our rooster, Chocolate, who is mentioned several times in these pages, began to see my three-year-old daughter as a threat. For a few months, I tried everything to get him to change his behavior toward her, to no avail. It became clear that I couldn't risk harm coming to my daughter, so I found him a home with a new flock and no small children.

THERE'S ONE IN EVERY BUNCH

YEARS AGO ON A SUNNY SPRING DAY, I went to the feed store on an ordinary trip. I didn't want any new chickens, but from the chick bins, their peeps called to me. I couldn't resist adding one sweet little ball of fuzz to the bunch of chicks that my Silkie hen Dolly had just hatched. I named the new girl Dottie Speckles.

The others took to her instantly, and they all grew and grew until it was time to move them into the big coop with the existing flock. Those girls were bigger, older, and wiser. Dottie Speckles spent plenty of time trying to hide under Dolly's wings. Dolly protected her despite the fact that her adopted daughter was now three times her size.

After a few months, I noticed that Dottie Speckles was starting to unleash her inner alpha hen. At first she attacked the Silkies, even Dolly. She pulled out feathers and chased those minding their own business, doing most of her malicious mischief at night. As the others slept, she took joy in plucking the feathers from their necks. Peck marks replaced feathers. She was becoming a bully.

I tried everything to get her to change her ways, but it eventually became clear that Dottie Speckles was going to need a new place, perhaps one with a rooster that could keep her in line. She wasn't interested in becoming head hen. She just wanted to terrorize her flockmates.

I rehomed her down the road at a nearby farm with plenty of chickens, including roosters. She did fine for a couple of weeks, and then she returned to her old ways. One night she plucked all the feathers from a Polish rooster's head! The other chickens spent their days hiding high up on the outside roosts. Her new family tried her in three different settings with three different groups, but none were successful. My friend told me that in 40 years of keeping chickens, she had never met a chicken with such a personality.

Dottie Speckles was placed in a smaller coop and run where she could see the other flocks but could no longer do any harm. Then one day a woman who wanted a single chicken for a pet visited the farm. She and Dottie Speckles soon happily left together for a new beginning.

Dottie Speckles

I don't really think Dottie Speckles realized what she was doing was wrong; that was just her. Perhaps there is a difference between bully hens and people who bully, in that people often learn the behavior from how they are raised, while hens are just born that way. Even so, with both people and hens, the ones at fault are not those who are bullied but rather those who are doing the bullying.

Overaggressive Mating

Protecting his hens and ensuring that they are laying fertile eggs is a rooster's job. Hens lay a lot of eggs, and roosters mate many times per day. They have favorite hens, and eventually it is not uncommon for those favorites to show signs of over-mating, such as missing feathers, claw injuries on their backs, and sometimes fatigue.

Even a hen who is receptive to a rooster's advances may try to escape from him if he tries to mate with her too much. A rooster should have about seven hens all to himself so that he does not spend too much attention on any one hen.

A hen with a bald patch from an overly attentive rooster may need to wear a chicken saddle while her feathers recover.

The Ultimate Protector . . .

Most chicken people probably have some story about a mean rooster. These encounters usually involve the rooster charging at or chasing them, sometimes with his spurs out. But I can testify that not all roosters are mean.

When people started breeding chickens to enhance particular traits, personality was usually left out. Many breeds have been developed solely for their ability to produce offspring or just for their beauty. In the past few decades, chicken advocates such as The Livestock Conservancy have made an effort to preserve

An aggressive rooster may injure other birds or even humans with his formidable spurs.

heritage breeds, including selecting for better-tempered roosters that can coexist peacefully with a human family.

One of the primary benefits of having a rooster in your flock is that no matter his size or breed, he takes his job as watchman and guardian seriously. His goal is to protect his hens from danger. He will literally give his life for his ladies. I know of at least two roosters that escorted their flocks to safety, then fought to the death against predators. Their bravery saved the lives of their hens.

Roosters stand outside the pecking order, which primarily exists for the hens. They help break up female disputes and may prevent other hens from picking on one particular hen. A rooster often has his favorite girl, with whom he spends most of his time. She is not necessarily at the top of the pecking order, but he will treat her like a queen. It's possible that other hens might be envious of her role, because when the rooster is removed, his favorite hen is sometimes picked on by the others.

. . . and Fearless Romantic

Not all hens will willingly mate with a rooster, even though he is the leader of the flock. With those reluctant hens, he must work to win their attention and affection. It is not always easy. One way roosters entice hens is through the dancing and vocalization patterns known as tidbitting (see page 20). The dance is wonderful to watch, and it is so much fun to see the rooster and hen interact.

My dear Oyster Cracker never gave in to our rooster's bid for affection. She always looked straight at him and told him, with pecks, to get lost. Other girls, however, could not get enough of his chivalrous ways!

Beauty is in the eye of the beholder.

HANDSOME IS AS HANDSOME DOES

Hens pay attention to several physical traits when assessing a rooster as a potential mate. In addition to noting the size of his spurs, comb, and wattles, hens are attracted to his eyes (specifically the iris, or colored part of the eye), his earlobes, and the feathers of his neck and tail. Size also seems to play a key role. Bigger roosters have more success. However, researchers have determined that it's not just the physical appearance of a rooster that

If a hen accepts the advances of one rooster, then mates with a higher-ranking one, she can eject up to 80 percent of the first rooster's sperm, increasing the chances that the second one will fertilize more of her eggs.

attracts females but also his dance moves. A rooster that makes robust tidbitting movements tends to attract the ladies' attention.

It also appears that the type of food a rooster uses while tidbitting directly correlates to his success. Desirable treats include corn or other favorite seeds, as well as worms, insects, and even small rodents, snakes, or frogs. When a rooster uses food to lure a female into mating, the hen evaluates the situation by listening and observing the male's actions; then she forms an opinion. In a flock with multiple roosters, each hen will watch all of them before expressing her mating preference. However, if one chooses to mate with a beta rooster, she will have to do it on the sly.

How Beta Roosters Win

How does a rooster low on the totem pole court hens without the alpha male noticing? In a flock with an alpha male and a number of secondary roosters, the beta roosters have been shown to use a surprisingly clever strategy: Thirty percent of the time, beta males tidbit silently, showing off for hens out of sight of the dominant rooster and without drawing his attention by vocalizing. While tidbitting silently, the inferior males monitor the alpha male's location and keep an eye on his interactions with the flock. They are equally successful at enticing females to mate with them despite the lack of "chicken speak."

HUMANS IN THE FLOCK

IT IS NOT UNCOMMON FOR FLOCKS OF CHICKENS to add their human caretakers into their ranks. One fall day, I was raking leaves and cleaning up the yard while the chickens were out free-ranging and exploring. Scratching and pecking in freshly fallen leaves is one of their all-time favorite activities.

Soon enough, it was time to move to a new spot in the yard. The chickens were still out, and because I had spotted a red-tailed hawk, I wanted them back in the safety of their run. Unfortunately, they decided to hide under the huge rhododendron bushes on the side of the garden. I could hear them calling to one another, announcing their finds, and could hear them scratching in the leaves, but I could not reach them.

As I was squatting by the bush, wondering how I was ever going to retrieve this motley crew, my neighbor came by to say hello. I walked over to the path that led from his yard to mine. To heck with the chickens, I thought. They'll come eventually.

My neighbor and I chatted for a bit. Then I noticed my rooster, Chocolate, down by my ankles, scratching and pecking and puffing up his chest. It took me a second, but then I realized that Chocolate was trying to tell my neighbor that I was his girl!

I scooped him up, gave him a nuzzle, and then set him down. He continued to stay by me. As we stood there, along came the rest of his flock, no coaxing necessary. That was the day I learned that I was seen as a member of the flock.

ARE YOU A HEN OR A ROOSTER?

When I first heard stories about hens becoming roosters, I didn't believe it. The stories were all the same: They started with a flock that either lost its rooster or never had a rooster. A hen in the flock would start crowing; molt and grow back rooster-like feathers; develop male sex characteristics, like a larger comb, a wattle, and spurs; and even try and jump onto the backs of the other hens in the flock. Could it be?

Although rare, changing sex can indeed happen.

Hens have only one functioning ovary; the other is typically undeveloped. An infection or a hormonal change in the functioning ovary, or even a tumor that increases testosterone production, can bring on an increase in androgen that causes the development of secondary sex characteristics. It can happen to a hen later in life and often isn't noticed by the chicken keeper until the hen's behavior changes.

In rare cases these hens may even produce semen, but most chickens that undergo a sex change do not produce offspring, although it can happen. The reverse change — male to female — has not been shown to occur.

FLOCK TALK

JUSTIN HEWITT AND STEPHEN THORNHILL

Backyard Chicken Keepers

Detroit, Michigan

Justin and Stephen own an award-winning app development firm. Active community members, they also restore historic homes, collect vintage cars, and travel as often as they can. Justin's love for all animals stems from having grown up participating in 4-H clubs and raising sheep, goats, and all types of poultry.

Current flock

A mixed flock of 20 that includes Australorps, Easter Eggers, Buff Orpingtons, Welsumers, Barred Rocks, and Rhode Island Reds.

What's special about your chicken coop?

We were told that the chicken coop has been there since the house was built in 1919, and that every owner had chickens at one time or another, so Justin remodeled the coop to more closely resemble the house.

Favorite chickens

Justin's was Mammy, a broody Silkie who once hatched out six ducklings. It was so funny to see them follow her around. She would get so upset when they went swimming!

Stephen's favorites were a pair of inseparable Buff Polish sisters named after Carol Channing and Phyllis Diller. They looked like they were wearing lopsided wigs that bounced when they ate — it was hysterical.

What do you love about keeping a flock?

We are always amused to see their individual personalities develop. The Polish and Silkies stick to themselves, but the little Silkies are scrappy while the Polish seem more concerned with being elegant and graceful. We enjoy letting them roam the yard after work while we watch "Chicken TV" and have happy hour. Plus, watching a group of baby chicks grow up never gets old.

What Makes a Chicken Tick?

Looking Past the Feathers

When we humans compare ourselves with other animals, those animals are most often our close relatives in the mammal world, especially chimpanzees. In many ways, however, we aren't all that different from chickens. Looking past the fact that we have arms instead of wings, we share many body parts and functions. In some ways we are superior to chickens, but in other ways they have us beat!

The chicken was the first bird to have its genetic composition mapped. When the project was completed in 2004, researchers reported that we humans share about 60 percent of our genes with chickens, compared to the 88 percent we share with rats.

AMAZING TRUE FACTS ABOUT CHICKENS

It's hard to outrun a chicken. Chickens can run as fast as 9 mph (14.5 kph), while humans average just over 8 mph (13 kph).

The average hen lays 265 eggs per year.

The Avam Cemani, a rare Indonesian breed, is entirely black, including feathers, skin, and even internal organs! The Silkie is also all black from its skin to its insides. This phenomenon is the result of a genetic condition known as fibromelanosis.

Chickens can't see in the dark.

The chicken is the closest living relative to the *Tyrannosaurus rex*.

Chickens don't sweat. They regulate body temperature through their combs and wattles.

The average chicken has 7,500 to 9,000 feathers, made of keratin, the same protein as in human hair and nails.

They may not have teeth, but their barbed tongues catch food and move it to the back of throat. Chickens produce saliva, but with few taste buds, they can't taste sweetness.

Most breeds of chickens have four toes, but some have five.

Chickenhearted: A chicken's heart beats some 400 times per minute, compared to the human average of 60 to 80.

You can tell what color egg a hen will lay by looking at the color of her earlobe. Red earlobes usually mean brown eggs. White earlobes mean white eggs.

ear (behind tuft of feathers)

ear lobe

EYES OF THE BEHOLDER

Like humans, chickens rely heavily on sight for survival. As a prey species, they have eyes on the sides of their heads, which increases their field of vision. Their brains take the sights from both eyes to create a 300-degree panoramic view without having to move their heads. Predators, including humans, usually have eyes in front of their heads, giving them binocular vision and a more limited peripheral field. Chickens have both monocular vision and binocular vision. They use monocular vision when they examine an object closely or watch for predators with one eye. Binocular vision helps with depth perception, which comes in helpful while scavenging.

Like us, chickens have an upper and a lower eyelid, but they also have an inner eyelid, or nictitating membrane, which folds into the corner of the eye near the beak. When a chicken blinks, the upper lid doesn't move like a human's. Instead, the inner eyelid slides over the eyeball to clean away debris while the lower lid

Like most birds, chickens have a nictitating membrane under the eyelids that cleans and protects the eye.

lifts to meet the upper lid. Another difference between human eyes and chickens eyes is that overall, chickens have better vision than we do. They are better at noticing differences between colors than humans are, and they can even see ultraviolet light.

A Bird's-Eye View

The really amazing thing about chicken eyes is that they contain a state of matter not seen in other animals. Called disordered hyperuniformity, it transmits light with the efficiency of a crystal but has the flexibility of a liquid.

The chicken eye has five types of cones, four that receive and interpret red, violet, blue, and green, and a fifth that detects changes in light levels. The cones, which are different sizes, are arranged in a single layer of tissue on the retina. Researchers used to assume that the cones in avian eyes were arranged in a repeated pattern, the same as mammalian ones. This was thought to be the optimal arrangement.

But the different-sized cones in birds' eyes do not touch and cannot be uniformly clustered together in a predictable pattern. Most birds have exceptionally keen vision, so clearly there is something to the lack of uniformity. "Nature found a unique workaround to the problem of cramming all those cones into the compact avian eye," according to Dr. Joseph Corbo of Washington University in St. Louis, Missouri, one of the first to study disordered hyperuniformity.

It also appears that chickens make specific and purposeful head movements, tilting up and down, forward and back, and side to side, to ensure that their entire eye, including the different types of cones, is able to process what they are seeing. It is almost as if they are using a scanner on a computer.

Do Chickens Have a Third Eye?

The pineal gland, located on the top dorsal surface of the brain, has been called the chicken's third eye because it has photoreceptors similar to the ones in the eye. This gland produces melatonin and regulates the circadian (daily) rhythms that dictate sleep cycles and other bodily functions. The pineal gland also triggers egg laying. We now realize that a chicken's pineal gland, once thought to react only to the amount of seasonal light changes, also reacts to changes in the type of light. As the seasons change, so does the spectrum of light that enters our eyes. Light affects the minds and body functions of the chicken, including egg laying.

Are you looking at me?

WHY DIDN'T THE CHICKEN *FLY* ACROSS THE ROAD?

Chickens aren't great flyers. They can fly about 6 feet in the air, managing spurts of about 20 feet in length. Their ancestors lived on the ground and spent their lives foraging on the jungle floor, not searching for food from the air. Their relatively long legs and feet were well adapted for walking and scratching, but their wings only needed to be strong enough to carry them into nearby trees to roost. Today's wild jungle fowl can fly a bit better than domesticated chickens, whose flying ability has mostly been bred out of them in favor of meat or egg production.

THE AMAZING BEAK

I always say that chickens peck first and ask questions later. The beak is to the chicken what hands are to humans. Chickens are curious, and they use their beaks to give them more insight into whatever object they are investigating. The chicken beak has a high concentration of sensory nerves and touch receptors. This multipurpose tool is used for gathering food (including catching prey), courting, raising young, fighting, preening, manipulating objects, and even navigation. My favorite time to see chickens use their beaks is when I give them a bowl of cooked spaghetti — it's so funny to see two of them squabble over the same noodle while each holding onto an end of it. I also love watching a broody hen ever so gently roll an egg away from a neighboring nesting box with her trusty beak to add it to her own clutch.

The beak is to the chicken what hands are to humans.

As Rare as Hen's Teeth

In 2006, by manipulating a chicken's DNA and activating a once-recessive gene, scientists were able to make chickens with teeth identical to their fossilized counterparts from millions of years ago. Under certain extremely rare genetic circumstances of gene expression, it is not impossible that chickens could develop teeth again. That ability has been maintained in their genetic composition.

The chick makes the initial crack with the point of the egg tooth, then uses the sharp sides of the tooth to finish breaking through the shell.

HOW DO CHICKENS SMELL?

To me, they smell fabulous. It is a mix of sweet hay, pine shavings, and soft, warm downy feathers like my pillow. Their smell is comforting and soothing. It evokes fond childhood memories of being on the farm with my best friend.

But seriously, chickens and humans have a similar number of olfactory receptors (scent sensors). Chickens use smell to find food and to sense when predators are near. Recent research on penguins allows us to infer that chickens might also possess the ability to use scent to distinguish individuals and identify ones they are genetically related to. This ability could help them to prevent inbreeding within the flock.

Chickens sneeze to clear nostrils that are clogged from inhaling dirt and food debris or from an infection.

WHAT DO CHICKENS DREAM ABOUT?

Most birds, including chickens, experience periods of rapid eye movement (REM) sleep, which is associated with dreaming. REM sleep occurs only in mammals and birds. During this phase, the brain is as active as it is while awake, but the body stays still. This is when dreams happen, but scientists so far have been unable to determine just what chickens might be dreaming about.

Chickens are very interesting sleepers. Because each eye is attached to the opposite side of the brain, chickens can be asleep and awake at the same time. If you ever try sneaking up on a

resting hen, you are likely to find her with one eye open and one eye closed. The portion of the brain attached to the closed eye is undergoing slow-wave sleep, which is the deepest type of sleep. The side of the brain connected to the open eye is fully alert and functioning.

This technique of sleeping is important during the day, as it allows for napping while staying alert for predators. At night, when chickens sleep in a row on a roost, the middle chickens can sleep with both eyes closed. The two chickens on the ends sleep with their outer eyes open, but move to a new spot on the roost during the night to rest both sides of their brain. I've noticed in my own coop that while hens higher up in the pecking order are often the first to chose their places on the roost, the flock does rotate their positions during the night.

When sleep deprived, chickens are able to simply enter slow-wave sleep, even forgoing their monocular sleep and closing both eyes to rest both sides of their brain. Unlike humans, who may need hours of extra sleep to feel refreshed, chickens can make up for lost sleep in just a few seconds of slow-wave sleep. Curiously, chickens don't show signs of sleep deprivation, according to the researchers. How do chickens become sleep deprived? From disturbances such as fireworks, barking dogs, predators, or by researchers keeping them up all night to study them!

ARE YOU MY MOTHER?

WHEN BABY CHICKS ARE BORN, they imprint on the first moving object that they see. If it's not their mother, it might be another chicken or animal or even a person. When I order chicks in the mail, I like to have the entire family present when we meet the new babies for the first time. I think this simple measure helps us to form deep lasting bonds with our flock over the years.

The first time we got baby chicks, I placed a small digital thermometer into the bottom of the brooder to be sure the temperature was just right. With the push of a button, you could see daily highs and lows and other trends in addition to the current temperature.

As the chicks began to grow up and explore the brooder, they became very interested in the thermometer. When I peeked in on them, I'd often notice that they had pecked at the buttons, changing the display screen. When they pecked, it would beep in response, encouraging them to peck more.

After a few days, I realized that the chicks were spending way too much time hanging out around the thermometer. They weren't eating as much and were even sleeping piled up around their new friend. It occurred to me that they were imprinting on the thermometer and thinking of it as a mother figure. When I removed the thermometer, the chicks resumed more normal behavior: milling about the brooder, exploring in the shavings, and practicing their roosting on small branches from the yard.

INNER WORKINGS OF THE EGG

From the time the egg is fertilized, the chick develops rapidly. It takes just 21 days for a chick to grow inside the egg and hatch. The eggshell is made of almost pure calcium carbonate, with more than 17,000 tiny pores to allow air and moisture to pass through. All eggs are coated in a bloom (or cuticle) that helps prevent bacteria and viruses from getting inside the egg. Here is a peek inside this incredible process.

Day 1 — The fertilized egg begins to develop the embryonic disk, where the embryo forms. An air sac in the rounded end adjusts internal egg pressure and air for respiration as the chick grows.

Day 2 — The vitelline membrane, which surrounds the vitellus (yolk), begins to form. It allows the albumen (egg white) to pass into the vitellus to nourish the embryo.

Day 3 — Blood vessels begin to form, and the head, torso, brain, and beating heart are visible.

Day 4 — The amniotic sac begins to develop.

Day 5 — The eye is visible, and the head curls toward the tail. Toes are forming.

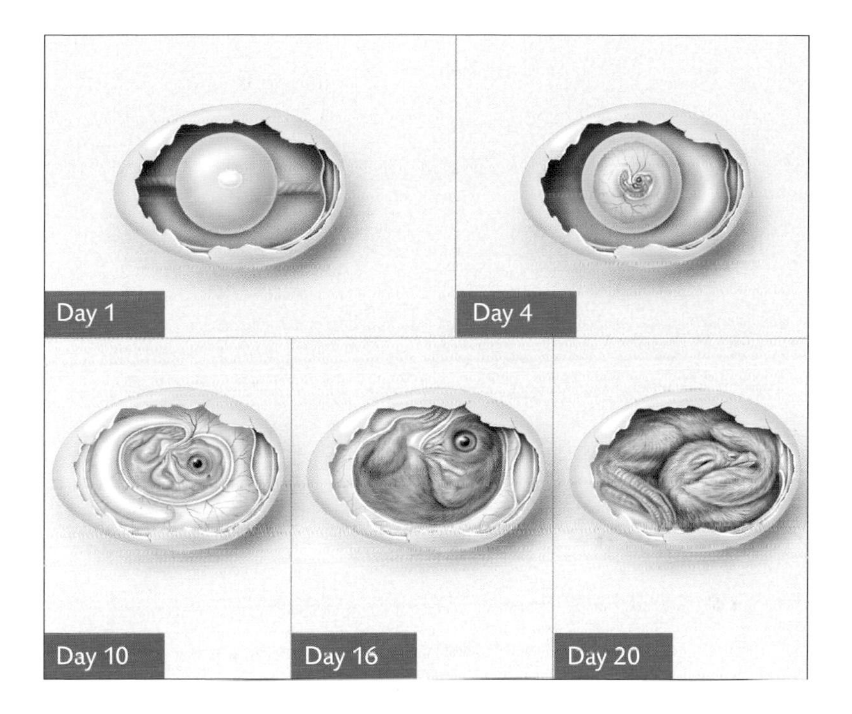

Day 6	The upper limbs grow. The vitelline membrane now encases half of the yolk.
Day 7	The beak begins to form.
Day 8	Eye color is discernible. The wings, legs, and beak are forming. The ear canals open, and the brain is set in place in the skull.

Day 9	Toenails and feathers begin to grow.
Day 10	The egg tooth appears and more feathers begin to emerge. The nostrils are visible and the eyelids have grown.
Day 11	The embryo clearly resembles a chick.
Day 12	The chick can hear noises from outside.
Day 13	Leg scales develop.
Day 14	Fluffy down covers the entire body.
Day 15	The yolk is shrinking rapidly. The chick's head tucks under the right wing to prepare for hatching.
Day 16	The albumen is almost completely consumed.
Day 17	The beak moves into the air space. The kidneys begin to function.
Day 18	Remaining fluids are absorbed.
Day 19	The beak is against the shell ready to pip, or pierce the shell.
Day 20	The yolk is fully absorbed and the umbilicus closes. The chick begins to pip.

Day 21 — The chick pecks around the entire circumference of the egg with its egg tooth and breaks through the shell, typically taking 12 to 18 hours with naps along the way.

FINALLY, THE CHICKEN OR THE EGG?

"Which came first, the chicken or the egg?" is a question posed since the days of the ancient Greeks. The answer was debated by the Romans, the Christian philosophers, Italian historians, and even Darwin himself. Modern scientists now agree that thousands of years ago, two chicken-like birds got together and produced an egg that hatched a bird that resembles the modern chicken. Because the protein that builds the eggshell is made by the hen, scientists have determined that it had to be the chicken that came before the chicken egg. Without the hen, there would be no egg.

A hen is only an egg's way of making another egg.

Richard Dawkins

NATURE VERSUS NURTURE

THE QUESTION OF WHAT LEADS CHICKENS to behave the way they do fascinates me. I think it is a little bit of both nature and nurture. Certainly the studies I have read, and my own observations from raising and keeping chickens, make me think that chickens are born inherently knowing how to be chickens. Those raised in the absence of a mother hen still learn all by themselves how to preen, scratch, eat, roost, and communicate during their first few weeks of life. We humans have nothing to do with that. Chickens are born with personalities, intuition, and intelligence to varying degrees; just like humans, they are all different.

However, chicks that are held often from the time they hatch to adulthood enjoy and sometimes prefer interactions with humans; this is especially true of certain breeds. And chickens are trainable. They can perform for us, and they can often make the most wonderful pets. Some people even keep house chickens. I think chickens are just like any other animal. It's a combination of nature and nurture that makes us who we are.

FLOCK TALK

KIM AND FRANK ROCHA

Backyard chicken keepers

San Antonio, Texas

Kim Rocha is the organizer of the San Antonio Backyard Pet Chicken group. She and her husband, Frank, enjoy sharing their knowledge, passion, and love of backyard chickens with others.

Current flock

Mostly Silkies and Frizzle Cochins, with a couple of Faverolles, a Wyandotte, a Polish, a couple of Araucanas, and two call ducks. Altogether, we have 20 feather babies.

Describe your chicken landscape.

We used to live downtown with our little flock, but now we have a big coop and plenty of space in the suburbs. Chickens can change the way you view life and what is important. If we had not started raising chickens, we would probably still be living downtown. I think of my life as BC and AC — Before Chickens and After Chickens.

Favorite chicken

There is always one that captures your heart. Snooki is the last remaining hen from our original flock. She is the first out of the coop to greet me. When I throw out treats, the others run toward the food, but Snooki always stays at my feet and waits for me to place her treats in front of her.

I love chickens because . . .

If they never laid another egg, they would still be the most fascinating creatures. They help me understand that if a little chicken can have such a beautiful soul and a distinct personality, then all animals and all people are special.

The best part of keeping chickens is meeting other chicken people!

Hey, I'm No Birdbrain!

Understanding Chicken Smarts

Calling someone a birdbrain is more of a compliment than it used to be. At least that's the way that I view it. The intelligence level of chickens is remarkable. Researchers have proven this, and chicken keepers know it all too well as they watch their flocks figure out how to sneak into off limit areas, learn where the treats are kept, and develop relationships with individual people and with other animals. The Internet is full of clever chickens showing off their tricks — maybe you could teach your own chickens a thing or two!

> *It is now clear that birds have cognitive capacities equivalent to those of mammals, even primates.*
>
> Dr. Lesley Rogers, author of *The Development of Brain and Behaviour in the Chicken*

BIRD BRAINS

For years, biologists assumed that creatures with larger brains must be more intelligent than those with smaller brains. It was believed that only mammalian brains had a forebrain and a neocortex. The neocortex forms the outer layer of the brain, where complex cognitive thoughts — problem solving, the creation of memories, abstract thinking, and sophisticated communication — are processed. The neocortex is also where personality and emotions are located.

Neuroscientist Harvey Karten of the University of California, San Diego, and his colleagues have sought to overturn this narrow-minded thinking for 40 years. By mapping the parts of the chicken brain that have to do with listening, they discovered that much like the human brain, the chicken brain forms columns of different types of interconnected cells, creating microcircuits between them. These arrangements are virtually identical to those found in mammals. It turns out that birds are much more capable of complex thought processes than was previously believed.

SOUNDS LIKE A MONKEY

When Dr. Chris Evans of Macquarie University in Australia would speak at animal behavior conferences about his research, he liked to list the following characteristics of his subjects:

- They are social animals that live in stable groups with an established hierarchy.

- They recognize one another as individuals and form different relationships with different members of the group.
- They have their own language and are able to communicate on a variety of topics.
- They adjust what they are saying based upon who is in their immediate vicinity and what is happening around them.
- They can solve problems, learn from experience, and teach one another.

Conference attendees often believed he was speaking about monkeys instead of chickens!

CHICKEN SEE, CHICKEN DO

Chickens and humans learn some things in the same ways. Like us, chickens are curious and they use their senses to explore and investigate. One way humans learn is by observing others and imitating their actions, and it turns out that chickens do this as well.

Dominant hens, and the head hen in particular, are usually the teachers within the flock. They are the ones who try new foods and introduce them to the flock. They explore more and are more likely to figure out puzzle-like equipment, such as critter-free feeders that require the chickens to step on a plate to open the lid to the food. When a dominant hen discovers something new, the others observe and copy her behaviors.

There are plenty of stories about house chickens that enjoy watching television. Chickens will sit for hours, sometimes watching and roosting on the back of the couch with their humans. What do chickens get from watching TV? It turns out that not only are they entertained, but they can learn things.

Hens can learn to modify their behavior from watching videos of other hens.

In one study, a group of chickens were shown videos depicting another flock in an outdoor space with two feeding stations in different locations. One station, featuring better feed and more treats, was located near shrubs, edible plantings, and other chicken-approved landscaping that provided optimal safety. The other feeding station was less inviting. The chickens in the video naturally showed a preference for the better station. When the chickens that watched the video were placed in the same environment, they appeared to recognize their surroundings and immediately sought out the better location with less exploring than chickens that didn't see the video.

In another study, hens watched a televised hen eating feed out of a red feeding bowl. They were then offered three red and three yellow feed bowls with the feed covered with sawdust so that color was the only clue. The hens that watched the video not only immediately chose to explore the red feeding bowls over the yellow ones, they continued to show a preference for the red feeding bowls in the future.

WHAT'S IN THE CAN?

I **KEEP CHICKEN TREATS** such as mealworms, sunflower seeds, and dried fruits in a small metal garbage can in a shed near the coop. Every morning, I stop at the shed before I start my chores to collect some treats in a container that I carry to the coop. I let the girls out to mill about in the run while I top off the feeders and waterers, harvest the eggs, and tidy up the coop. Once I'm done, I hand out the treats.

The girls recognize the container. As soon as one chicken sees it, she alerts all the others. Once I have their attention, I call out a singsong, "Giiiirrrrrlllllssss!" and they get giddy with excitement.

One day when the girls were out free-ranging, they happened upon the open shed and saw the garbage can inside. Curious, they looked at their reflections in the shiny surface, and then one hopped up on the lid and began rubbing her beak on it. It's possible that they could smell the treats in the sealed metal can, but I think they inferred what was inside it from having seen me emerge from the shed with the treat container. I didn't have the treat container. I didn't call out to them. Yet somehow they knew. They started dancing as I lifted the lid. How could I not reward them?

IT'S SPRING!

I **HAVE ALWAYS WONDERED ABOUT THE FLOCK'S ABILITY** to remember different things. It seems they retain memories that matter the most to them while discarding more trivial events.

For example, when I first start the lawnmower in the spring, the chickens know exactly what comes next despite months passing since they last heard that sound. From the run, they watch me mow, dancing in anticipation of the fresh clippings. As I approach, they leap with excitement. After I toss the clumps of green goodness to them, they chortle and scratch in the clippings as if they have been waiting all winter.

And yet, every October I place a few pumpkins in the run, and every year, they act like they're seeing aliens. They carefully avoid the strange orange orbs for few days. Finally, after one brave soul takes the first peck, they all dive in to devour the treat, enjoying the seeds the most!

Why do they seem to forget the pumpkins and not the grass? Do their brains retain the grass clippings because mowing happens regularly over several months, unlike the once-a-year pumpkins? Do they "delete" the pumpkins because their brains can only hold so much? It's fascinating, indeed, and I am always learning more!

BRAIN FUNCTION

Bird brains are small in comparison to many mammal ones, but scientists have discovered that birds are able to process complex information because their brains are more densely packed with neurons than those of some animals to which they have been compared. One of the most fascinating things about bird brains and human brains is their ability to regenerate neurons, or nerve cells. However, the causes of regeneration are markedly different. Humans typically regenerate nerve cells after an injury, such as a stroke. This is called regenerative neurogenesis. Avian brains, on the other hand, can regenerate neurons without sustaining an injury. This ability allows birds to take apart and rebuild portions of their brain based on their current needs. Scientists believe that if they are able to replicate this unique ability of bird brains, then they may be able to cure depression, Alzheimer's disease, and other conditions that affect the human brain. In this situation, who wouldn't want to have a bird brain?

Here's how it works. The parts of the brain that hold information that is constantly in use are allocated a disproportionately high number of neurons compared to other parts of the brain. The parts of the brain that store information used less frequently can be downsized. The parts that store information that goes unused can be ramped up with more neurons later when needed or simply deleted. Because the overall size of the brain does not change inside the skull, when one portion of the brain increases in terms of the number of neurons, another portion decreases.

When the bird creates new neurons, they are used by the part of the brain that the bird is currently using the most during that given period of time An example is the natural cycle of neurons in the portion of the brain that holds breeding songs. In the spring, seasonal hormonal changes trigger the brain to amp up the area where songs are located, allowing males to sing their best songs to attract mates. Those areas shrink later in the spring when breeding songs are no longer required.

With this ability to erase and replace the neurons in their brains, chickens are able to access information that they need to know, and then let it go if it becomes unimportant, while rebuilding sections of their brain to store new, more relevant information. Chickens use their memory-forming and storage ability to keep track of flock relations, including the pecking order and mating preferences. They can recognize as many as 100 other chickens, as well as remember individual people and other animals, both friend and foe. They recall the location of favorite food spots when free-ranging and quickly learn to associate certain activities or containers with feeding time or the handing out of treats.

PATIENCE IS A VIRTUE

In a study at Silsoe Research Institute in England, chickens were shown to have the ability to delay gratification by putting off an immediate but minor benefit in preference of a later but larger payoff. For the first part of the study, hens were trained to push buttons that resulted in the opening of a door that allowed access to feed. Once the chickens realized that they could access the feed by tapping on certain buttons, the researchers explored the concept of delayed gratification: Would a chicken be patient enough to wait longer in order to receive more feed?

Researchers programmed the buttons so that if a hen pecked on one button, she had to wait 2 seconds and then the door to the chicken feed would remain open for 3 seconds. A second button offered a 6-second delay rewarded with 7 seconds of feeding time. Only 22 percent of the hens in this phase of the study showed a preference for the button that gave the longer feed time, most likely because an additional 4 seconds of access wasn't interpreted as a superior choice.

In the next phase, a second set of hens were trained to use the buttons, but this time a "jackpot" option was added, in that the 6-second delay button gave the hen access to the feed for a full 22 seconds instead of a mere 7 seconds. In this phase, 93 percent of the chickens chose to wait and "hit the jackpot" rather than go for the more immediate reward. The chickens learned that a longer wait resulted in a noticeably bigger payoff.

I have seen a type of delayed gratification response in my own flock. Each morning before I let the hens out of the coop, I scatter a couple handfuls of a scratch and a handful of mealworms in the run. The chickens adore mealworms and seek them out, just like little kids who pick the marshmallows out of their breakfast cereal.

One day, I left a new bag of mealworms in the house, so I tossed only scratch into the run before going to get it. Do you know that the entire flock ran around the run looking for those mealworms among the scratch? They would not eat the scratch and called to me as if to say, "Hey! You forgot something!" They preferred to wait for the mealworms before eating the scratch. On occasions when I've been completely out of mealworms, I've seen them wait up to 20 minutes before realizing that there are no mealworms that day. Talk about chicken mom guilt!

Learning from Mama

A type of delayed gratification can be seen when a hen first teaches her newly hatched chicks about finding food. She will take them to the food, indicate it with her beak, and wait for them to discover it. Only after the chicks have figured it out and have had a chance to fill their tummies will their mother begin to peck alongside them.

Mothers also teach their babies which food is good to eat. In one study, potential mama hens were fed only food that was colored red. The hens recognized that the red food was safe to eat. Then researchers introduced a bowl of blue food to one of these red-food-eating hens and her babies. When the baby chicks began to eat the blue food, the mama hen began to scratch, peck, and vocalize an alarm, letting her chicks know that she perceived potential danger in the blue food. It was a simple experiment, but the glimpse into the mind of a mama hen was priceless.

CHICKEN MATH

Chickens are capable of rudimentary math skills, such as distinguishing smaller and larger numbers. In one study, groups of day-old chicks were reared with small plastic eggs. The researchers believed that the chicks bonded with these plastic eggs as they would a mother. So naturally, the chicks preferred to be close to the plastic eggs. This study consisted of two parts when the chicks were just three and four days old.

In part one of the study, two opaque screens were set up in the brooder and individual chicks were placed behind a clear piece of plastic where they could view the research team manipulating the eggs. The chick looked on as the researchers moved the plastic eggs behind the opaque screens. Once all the eggs were hidden, the chick was released to explore. The chicks consistently chose to peer first behind the screen with the most eggs, even when the difference was as little as one egg. The chicks always preferred the larger cluster of plastic eggs.

In part two, with the same set up, the chicks watched as the researchers moved the eggs back and forth between the opaque screens. Once again, when set loose to explore, the chicks gravitated to the screen that hid the higher number of eggs. The chicks were able to keep track of the eggs and how the numbers changed between screens.

And Geometry

Chickens can also easily and quickly distinguish among geometric shapes. Clicker trained to peck at one of four different shapes, such as a circle, square, triangle, and rectangle, chickens in one study could always pick their shape out of the grouping, no matter how the shapes were arranged. When their particular shape was removed, the chickens looked quizzically for it and wouldn't peck at the other shapes. When the correct shape was reintroduced, they pecked at it as taught. Chickens are also able to distinguish a particular color out of assorted colored circles.

In another study, five-day-old chicks were presented with 10 shapes in a row. One of the shapes held a small bit of feed. When the shapes were reordered, the chicks still chose the shape with feed.

And Even Physics

Chickens use physics to predict where leaping bugs like grasshoppers will land. Instead of catching them midair, chickens project where the bugs will land and run to the estimated place of impact. Amazingly, they are usually spot on!

In one experiment, chickens presented with images that defied the laws of physics and images that fit the laws of physics looked longer at the accurate images than ones that were distorted. They also showed more interest in photos of real structures than abstract ones.

I'VE GOT MY EYE ON YOU

A number of popular YouTube videos and at least one television commercial show people moving chickens in circles to demonstrate how their heads remain perfectly still no matter what their bodies are doing. Whether the chicken's body moves up, down, sideways, or around and around, its head stays in the same place. How do they do it? By constantly and reflexively making precise real-time adjustments that counteract the movement of the body to keep the head locked in position.

Chickens lack the ability to move their eyes in their sockets and remain focused on an object as people do. Humans are able to keep their eyes focused on an object while moving their heads using their vestibulo-ocular reflex (VOR). This system connects the inner ear and the eye, allowing us to keep our eyes steady as our bodies move. Chickens, with ears anatomically similar to people, have the VOR but mostly rely on their vestibulocollic (VCR) and cervicocollic (CCR) reflexes.

Simply put, the VCR allows the neck muscles to compensate when head movement is detected by the ear. The CCR involves compensation by neck muscles to adjust for the neck's ability to tell the chicken's brain where it is in space and time (proprioception). Together these reflexes allow the chicken to remain oriented in space at all times by making fine adjustments with their muscles.

PROBLEM SOLVING AND DECISION MAKING

Scientific research is all well and good, but anyone who keeps chickens knows what they are capable of. They are full of surprises, and once they've figured something out, it's amazing to watch them adapt and improve on their techniques and abilities, such as taking a new route to reach the rooftop or using a bucket or woodpile to help them scale the fence around the vegetable garden.

- A friend surrounded her tempting garden with a high chicken-wire fence, then clipped her chickens' wings to prevent them from flying over it. Yet every day, her free-ranging flock somehow ended up in the garden. Hiding nearby, she watched to find out their secret. It didn't take long for one of the girls to hop up onto the fence, which sagged a bit toward the ground. Then another hen hopped up, and another. When the fence was low enough from their combined weight, they could reach the off-limits vegetables and plants. Problem solved!
- Chickens do better in cooler weather, so they seek out shade during warm weather. I've heard of chickens that come to the door looking to enter the house when they know the air-conditioning is on. Once inside, they not only cool down but also sneak a snack from the dog and cat bowls when no one is looking.

- Another friend has a small barnyard flock that roams freely, but when it is time for the kids to get off the school bus, those chickens go down to the bus stop like clockwork right before the bus arrives. They wait for the kids to get off and then walk them home to partake in after-school treats.
- I have heard of roosters informing chicken owners as they lock up the flock for the night that some of the girls are still outside the coop. Sure enough, a stray hen will be discovered in the garden. One rooster even knocked on the sliding glass door of the house to notify his human that danger was in the yard and the flock needed help.

The chickens were pecking feverishly at some weeds through the wire of the run. Molly simply walked out the open run door where she could enjoy a feast as the others watched.

TRICK CHICKENS

With all their smarts, chickens are highly trainable. They can be motivated with treats and clickers. They can easily learn to recognize patterns, shapes, colors, and images of other animals. They can be trained to follow and peck at a moving target. Here are some of the things chickens can do:

- Slalom through poles
- Jump hurdles
- Climb ramps
- Hop through hoops
- Run through tunnels
- Play the piano, xylophone, and drums
- Dance along to music while keeping the rhythm

Training chickens is so fun and rewarding that there are even camps where you can take your favorite chicken and teach her some new tricks!

Are you trying to visualize a chicken dancing to music? Picture her scratching at the ground with her head held high!

I have often found flock members standing outside the kitchen door begging for food!

DON'T BET AGAINST THE CHICKEN

Playing tic-tac-toe against a chicken might seem like a sure bet, but don't be so sure! For years, many carnivals and game arcades allowed human players to match wits against a trained chicken, which just about always won at this ancient game. The practice went the way of many such acts, but a chicken named Ginger took Las Vegas by storm in 2001. If you could beat Ginger, you could win $10,000! Ginger (there were actually 15 hens with the same name) played up to 500 games per day. In 135,000 games, Ginger only lost 5.

How did Ginger win so often? One trick was that Ginger always went first, which is a big advantage in tic-tac-toe. Also, while chickens are easily trained to peck at a gameboard for a reward, they aren't really playing the game. That's being done by a hidden computer.

THE CHICKEN'S COMPASS

So how do chickens find their way around? Most scientists believe that chickens use information from the sun and their ability to detect the magnetic fields of the earth to navigate. Researcher Cordula Mora, while a fellow at the University of North Carolina, Chapel Hill in 2004, posited that birds have a "compass" in their eyes and a "map" in their beaks. Chickens, starting as early as two weeks of age, use the sun's position to interpret their location. Birds have specific photoreceptors in their eyes that, when activated, release chemicals in the bird's brain that mark a particular location, such as a migratory destination, in the bird's memory.

Chickens also have a magnetic compass located in the base of their beaks between their eyes. Scientists believe that this compass comes from ancestors of modern birds that lived 95 million years ago in thick, lush jungles with little access to the sun. Despite not roaming far from home these days, chickens are able to locate their favorite places, foraging hot spots, hidden nests of eggs, and prime trees for roosting.

Not all who wander are lost.

J. R. R. Tolkien

FLOCK TALK

JENNY CARCIA

Artist and proprietor of Pet Chicken Ranch

Petaluma, California

Jenny combines her love of art, quilting, and chickens by making customized chicken models from beautiful fabrics.

Current flock

From a flock of six Easter Eggers, Barred Rocks, and Buff Orpingtons, I now have just Buffy, a glorious Buff Orpington who's always been head of the flock, and Latka, an angelic little blond Easter Egger, but we'll be adding more soon.

Favorite chicken

I shared a very special bond of unconditional love with Pork Chop. I experienced just about every emotion with that little Easter Egger. Sometimes she felt like my child, and sometimes like a best friend. I could always depend on her to make me feel better about a bad day.

Never "just chickens"

As an animal lover, I never thought of them as "just chickens," but I had no idea the vast measure of love I would feel for my flock. I'm fascinated how the relationship builds over the years. I could watch them for hours.

Lessons from the flock

Keeping chickens has made me more patient and more accepting with life in general. Chickens seem to live in the moment, and every moment seems full of lessons that teach me how to be a better human.

The best part of keeping chickens?

Their companionship. They are such dear little creatures. I love having them alongside me in the garden and just relaxing with them while they do their chicken things, from dust bathing to napping in the grass. I love the daily routines that we share.

How Do You Feel?

The Emotional Life of Chickens

When we got our first backyard flock in 2010, it was a complete leap of faith for me and my family. We had never kept chickens and had no idea what was in store. Some people want chickens mainly for the eggs. We welcomed the eggs, but we were more interested in the chickens as pets and in the life lessons they could teach us. It didn't take long for us to discover how wonderful keeping a flock could be. Perhaps what intrigued me the most was coming to realize that chickens have emotions, just like dogs and cats. I'm embarrassed to admit it, but I was somewhat astounded.

I saved this section of the book for last, as it is nearest and dearest to my heart. Call me a crazy chicken lady, but I have fallen in love with some of these girls. I like to think they love me, too.

They have become members of the family. When my family members and I are with them, our problems melt away. They make us laugh. They are silly, curious, and show moments of kindness. They exhibit compassion to their flockmates, worry when others wander off too far, sleep together at night, accept those from different breeds, have tolerance for differences, live with one another's unique quirks and personalities, and are amazingly resilient. I cannot imagine our lives without these wonderful companions in the yard and garden.

RULES OF ATTRACTION AND JEALOUSY

Numerous social science studies are devoted to how and why people choose their mates in life. I'm fascinated that our degrees of attraction vary so immensely, and was intrigued to realize that the same goes for hens and roosters. Roosters do not mate evenly with all the available hens. They have their favorites, with whom they mate whenever they get the chance. And not all hens agree to mate with the rooster. It went that way for Oyster Cracker. Despite his best showstoppers, our rooster, Chocolate, never got to mate with her. Now Dolly was another story.

Dolly was a beautiful gray Silkie hen who seemed to be perpetually broody. Perhaps that was what attracted Chocolate, because Dolly quickly became his girl. They went everywhere and did everything together. He would tidbit just for her, dancing and sharing seeds, worms, and slugs with her. He even presented her with frogs and once, a small snake. He would proudly parade her around the run among the flock. It was intriguing to see how the entire flock, even Tilly, the head hen, yielded to Chocolate and Dolly as they paraded along.

After we had to rehome Chocolate, it soon became apparent that Tilly was jealous of Dolly. With her protector gone, Dolly lost her privileges. Tilly placed her at the bottom of the pecking order, keeping her away from the feed, forbidding her from entering the coop at dusk until the very last minute, and kicking her out of the favorite nesting boxes. Eventually, Dolly seemed to just accept it and there were no more squabbles. Dolly remained in her new pecking-order place for the rest of her life.

FRIENDS FOREVER

OUR FIRST PEEPING PACKAGE FROM THE POST OFFICE included two golden Buff Orpington chicks. Classic puffs of yellow fluff, Oyster Cracker and Sunshine were inseparable from day one. They went everywhere side by side for their whole lives. They ate together, dust-bathed together, free-ranged together, and slept wing to wing on the same roost. They would even clean each other's beaks of leftovers after feeding.

They were also inseparable when it came to laying eggs, though they rarely laid their eggs at the same time. With a few hours between eggs, it was a hoot to watch them sort it out. Whenever one went into a nest box, the other would stay nearby, clucking encouragement. The observer would pop out to explore the run briefly, then return to the coop to check on her friend, peering into the nesting box. Back and forth she went until the egg was laid. It wasn't unusual for them to sing the egg song together, no matter which one actually laid the egg.

After we lost Sunshine, it was hard to watch Oyster Cracker call out for her best friend day after day. I tried to ease her pain by spending time with her. When I went into the coop, she would hop into my lap, and often I'd have to scoop her off when I left. We had another six months together, but she was never quite herself. She seemed withdrawn, and even though she eventually warmed up to Feathers, one of our Silkie hens, it was not enough. When she passed, I believe it was partly from a broken heart.

Oyster Cracker and Sunshine

None of our other hens have ever taken the same degree of liking to one another as did these two lovely ladies. Some friendships are once in a lifetime.

Friendship is a single soul
dwelling in two bodies.
Aristotle

COMPASSION

I have witnessed compassion from hens toward one another, specifically when a flock member is injured or ailing, or if they can't find a member of the flock. They show their concern by visiting and trying to coax the ailing hen along. They call out for those that are missing. They cock their heads like dogs do, to listen and try to understand the situation.

The compassion of chickens to those outside their flock can be pretty incredible, too. Chickens are increasingly being used as therapy pets. Some nursing homes keep flocks as a way to engage elderly patients, many of whom love to watch them or help take care of them. The hens may perch in residents' laps or on top of their walkers and just sit quietly. Some chickens fall asleep, while others are content to observe the world around them. Having the birds around gives the residents something to talk about and focus on. The chickens have a special way of keeping them interested and making them feel needed.

In a strange way, chickens seem to know when you are feeling down or depressed. Spending time with the flock always seems to melt those feelings away. A few years ago, my son went through a very difficult time with bullying at school. He was only nine, but I could see how being bullied was taking over his world. On particularly difficult days, he would come home from school and forgo his afternoon snack to immediately go see the girls. He

would hang out with them and tell them what was on his mind. They kept his secrets, and he told me that they made him feel better. I believe the chickens saved him that year and for that I'm grateful.

Try to be a rainbow in someone else's cloud.
Maya Angelou

EXCITEMENT

One of my favorite things about living with chickens is that they have a hard time containing their excitement. No matter what the weather is, how they slept during the night, or what went on the day before, they greet each morning with a lust for life. Like little kernels of popcorn, they emerge through the pop door ready to spring into the day, full of energy and interested in everything. Just the prospect of a new day is enough for the chickens to become almost giddy with excitement.

Of course, chickens get excited over specific things, too. My flock gets excited about me and the family. They come running when they see us. They hang out with us in the yard. When I dig holes for new plants, they happily hop inside to take a dust bath or scratch at the dirt in search of a tasty worm or grub. An offering of fresh treats from the garden makes them squawk and chortle with anticipation and delight. Often, I'm not sure who is having more fun, me or them!

FUN AND GAMES

Chickens actively seek out happiness and pleasure. Their lives revolve around feeling good. Some chickens are rebels and thrill seekers. These are the ones that dare to cross the road even when no other chickens will follow. These are the ones that will meet you at the door, step inside your home, try to get over the

Every day may not be good, but there is good in every day.

garden fence, and even hide a nest full of eggs far away from the coop. Some chickens even choose their human flock over their chicken flock.

Sometimes chickens become bored and frustrated. This is especially true in winter, when they may not be able to free-range and their access to fresh green foliage and grass is cut off. Just as with humans, being literally "cooped up" can lead to naughty behavior to pass the time. With chickens, this can include ganging up on a lower-status hen, sometimes with relentless pecking, feather pulling, and drawing blood.

During these times of boredom and aggression, distractions can often provide a solution. When they have something fun to do, their behavior improves! Chickens will investigate and amuse themselves with just about any object in their environment. They adore investigating new things, they quickly learn to perform tricks, and they enjoy staring at themselves in a mirror. It never ceases to amaze me how inventive chicken keepers can be in providing entertainment for their flocks. From offering them veggie piñatas of cabbage or cauliflower to peck at, to building them climbing structures and swings, we can enrich their lives.

My girls love it when I rake up big piles of leaves in the fall. They scratch around and disappear into the heap as they seek out bugs and hidden mealworms that I toss out for a chicken treasure hunt. They also do silly things that only they understand, like the time Tilly found a small stick covered with fresh leaves and decided to drag it all over the yard. It became almost like her friend. She dragged it everywhere that day, and just when I thought that she had abandoned it, something clicked and she ran back after it to drag it along on the day's next adventure.

I dream of a better world where chickens can cross the road without their motives being questioned.

Unknown humorist

TRAUMA

Chickens react to horrible experiences the way many other animals do: They can sink into a depression and show signs of fear and distress for a long time afterward. I saw this in a friend's flock after a weasel snuck in and killed nearly all of her two dozen hens. Only three of them escaped to the rafters. One had a bite wound on her ankle and was visibly shaken.

Once brought to safety, the three survivors seemed to suffer some sort of post-traumatic stress. They stopped laying eggs and spent their days hiding. My friend spoiled them with fresh vegetables and added new roosts and other distractions. It took months, but eventually the traumatized hens began behaving more normally and were able to become part of a new flock.

Studies have shown that chickens recover more quickly from stressful situations when in the company of friends. They associate being with their flock as being safe and prefer to be with familiar chickens over unknown ones. With familiar companions, every chicken knows where she stands. With unknown chickens, rankings in the pecking order must be established, which is stressful for them.

MOURNING AND LOSS

As a hen nears the end of her natural life, she often goes off and finds a quiet place away from the rest of the flock. During this time, the other members of the flock visit, one by one or a couple at a time. To me, they seem sad. They "chat" through verbalizations and body language. They hang their heads low to get on eye level with the dying chicken. Their coos are quiet, soft mutterings that you have to lean into to hear — chicken whispers.

They move slowly and cautiously, considerate of disrupting their flock member. Some linger, while others putter around. Some sit with the dying chicken. Some keep returning. I watched some hens try to motivate Tilly by carefully scratching in the coop's pine shavings right in front of her, as if to say "Get up, please."

However, once they have made their peace, they leave and do not turn back. The dying chicken passes alone. The others have returned to doing what the flock does: foraging, scratching, dust bathing, and such. Life goes on.

Still, for days after a hen dies, it is not uncommon for those who were closest to her to mourn the loss of their friend. From the safety of the coop, they call out, using the same sound that means "Where are you?" when they are free-ranging in the yard and can't find a missing member of the flock. A grieving hen avoids interacting with the flock and sits in a corner with puffed-up feathers like a chicken that feels ill.

Some mourn only temporarily, but others never seem to recover from the loss of a flockmate. It is not uncommon for a hen that was close to the departed to suddenly pass unexpectedly, for reasons unknown. She seems to slowly lose her zest for life. Often, of course, she and the deceased hen were close in age and grew up together. I believe chickens can indeed die from broken hearts or, more specifically, from the stress and depression that follow a loss.

SAYING GOODBYE TO TILLY

I KNEW TILLY'S FINAL DAY WAS COMING. She had been slowing down, not herself but not overtly sick. One evening, when she opted for a pile of clean shavings on the ground instead of roosting, I knew the end was close. She held on until morning. I am convinced that she waited for me to say goodbye. I scooped her into my arms. She was almost lifeless, but she slowly opened her eyes. I held her, loved on her, and spoke to her in the gentlest of words. Mostly I thanked her. Thanked her for being such a wonderful hen. She taught us about love, friendship, the importance of rules, patience, and kindness, to be lighthearted and to laugh at ourselves.

When I buried her under a beautiful hosta, I placed a small bouquet picked from her favorite spots in my garden on her feathered wing, still warm from the sun. She had always loved sunbathing. I took a feather and tucked it close to my heart. I was numb. She was gone.

There are more than these little bits of her than in my heart. She is in my gardens, my children's memories, and mine, too. I can still hear her voice when I close my eyes. Tilly changed me forever. To this spunky little hen I will be forever grateful.

LOVE

I know that love exists between chickens. It's clearly evident that they care about one another's well-being and can form special bonds. Anyone who's ever loved a chicken knows that they form special bonds with non-chickens as well. I've seen hens sheltering a litter of kittens under their downy fluff and dogs snuggling with baby chicks. Chickens will perch on the backs of goats, and even horses, for a higher perspective.

I recently read about one special hen named Monique and her human companion, a French sailor, who are on the adventure of a lifetime, exploring the world in a 39-foot sailboat. She is the young man's only companion. She eats with him, sits on the boom while sailing, and pecks at the computer as he writes. They have gone sledding, windsurfing, and paddle boarding together. She knows her name, comes when called, and even shares a pillow with him when they sleep. From her original purpose of providing eggs, she has become a kindred spirit and fellow adventurer.

I've never taken them sailing, but I love my flock as I would any other pets. Of course, I do have my favorite chickens that seem to adore me as much as I adore them. Some come for chicken hugs. If I squat down, they come right between my legs and nuzzle into my neck. I feel tiny puffs of breath from their nostrils on my neck and behind my ear. They wriggle right in as if they can't get close enough. They look for me and call for me, using my chicken name.

I never expected that I would be friends with a chicken, but that is the wonderful thing about life's journey. If you take the time to explore something new in this world, often the unexpected happens. The most important part of the journey are the people and animals we meet. Our relationships ground us, provide cherished memories, and help us grow. Love is a universal language, and anyone who loves chickens knows that they speak it, too.

Until one has loved an animal, a part
of one's soul remains unawakened.
Anatole France

FOR FURTHER READING

Space prohibits listing all the sources I've consulted over the years. For a comprehensive list, please visit tillysnest.com.

Books

Nicol, Christine J. *The Behavioural Biology of Chickens*. Wallingford, UK: CABI, 2015.

Rogers, Lesley J. *The Development of Brain and Behaviour in the Chicken*. Wallingford, UK: CABI, 1995.

Rollin, Bernard. *Farm Animal Welfare*. Ames: Iowa State University Press, 2003.

Sewell, Franklane L., and Ida E. Tilson. *The Poultry Herald Manual: A Guide to Successful Poultry Keeping*. Saint Paul, MN: Webb Publishing Company, 1898.

Articles

Abeysinghe, S. M., et al. "Can Domestic Fowl, *Gallus gallus domesticus*, Show Self-Control?" *Animal Behaviour* 70, no. 1 (July 2005): 1–11.

Adler, J., and A. Lawler. "How the Chicken Conquered the World." *Smithsonian Magazine* (June 2012).

Edgar, J. L., et al. "Avian maternal response to chick distress." *Proceedings of the Royal Society/ Biological Science* (March, 2011).

Harms, W. "Smells may help birds identify their relatives." *UChicago News* (September, 2011).

Highfield, R. "So Who Are You Calling Birdbrain? Chatter of Chickens Proves They Are Brighter than We Thought." *The Telegraph* (November 15, 2006).

Kaplan, G., and L. J. Rogers. "Bird Brain? It May Be a Compliment." *Cerebrum* (April 1, 2005).

Rugani, R., et al. "Discrimination of Small Numerosities in Young Chicks." *Journal of Experimental Psychology: Animal Behavior Processes* 34, no. 3 (July 2008).

Smith, C. L., and C. S. Evans. "Multimodal signaling in fowl, *Gallus gallus*." *Journal of Experimental Biology* 211 (2008): 2052–57.

Smith, C. L., and S. L. Zielinski. "The Startling Intelligence of the Common Chicken." *Scientific American* (February 1, 2014).

Wolchover, N. "A Bird's-Eye View of Nature's Hidden Order." *Quanta Magazine* (July, 2016).

ACKNOWLEDGMENTS

Thank you so much to the amazing chicken scientists proving every day how similar we are to chickens. It is through your visions, hypotheses, and dedication that we continue to learn more and more about our remarkable feathered friends. Thank you to my friends and family, who have always believed in me and helped me make my dreams come true. I am blessed to have you all in my life. I love you. Thank you to my amazing team at Storey Publishing, including my editors, photographers, designers, and publicists. Finally, thanks to the fans and others who have been a part of my blog, *Tilly's Nest*, and whom I have gotten to know, love, appreciate, and adore over the years. I am glad to know that many of you feel the exact same way I do about our flocks!

IMAGE CREDITS

Additional images by © Aleksandra Mihok/EyeEm/Getty Images, 62; © Anthony Lee/Getty Images, 50, 72; © artpipi/Getty Images, 84; © Ben Queenborough/Getty Images, 100; © Bob Elsdale/Getty Images, 36; © Brian Summers/Getty Images, 25; © BŽatrice Schuler/EyeEm/Getty Images, 31; © Caroline Sale/Getty Images, 111; © Chesh/Alamy Stock Photo, 89; © Christian Michaels/Getty Images, 136; © Cristian Calderon/EyeEm/Getty Images, 44; © Cyril Ruoso/Getty Images, 68; © Daniel Kaesler/EyeEm/Getty Images, 14–15; © Dinodia Photos/Alamy Stock Photo, 51 left; © Dorling Kindersley/Getty Images, 16–17, 18; © Edward Kinsman/Getty Images, 95; © Ezra Bailey/Getty Images, 121; © FLPA/Paul Sawer/age footstock, 60 top; © Fuse/Getty Images, 131; © Gandee Vasan/Getty Images, 41 bottom; © Global IP/iStockphoto.com, 41 top, 88; © Hemis/Alamy Stock Photo, 51 right; © Image Source/Getty Images, 2–3, 4; © Ingram Publishing/Alamy Stock Photo, 32; © INTERFOTO/Alamy Stock Photo, 52; © Isabelle Plasschaert/Getty Images, 60 bottom; © Itsik Marom/Alamy Stock Photo, 69; © IvonneW/iStockphoto.com, 34; © Jamie Grill/Getty Images, 119; © Jane Burton/Getty Images, 53, 59, 113; © Jason Nastaszewski/EyeEm/Getty Images, 21; © Jeremy Maude/Getty Images, 19, 39; © JoeLena/iStockphoto.com, 80; © Jozef Culàk/iStockphoto.com, 71; © Juniors Bildarchiv GmbH/Alamy Stock Photo, 42; © Jutta Klee/Getty Images, 1; ©Keller + Keller Photography, 135; Kim Rocha, 99; © Laurie O'Keefe/Getty Images, 93; © Lee Grainger/EyeEm/Getty Images, 133; © Lena Granefelt/Getty Images, 48; © Life On White/Getty Images, 12, 35; © LM Photo/Getty Images, 30; © Lynn Stone/Getty Images, 23; © M-Reinhardt/iStockphoto.com, 85; © Maarigard/Getty Images, 38; © malerapaso/Getty Images, 40; © Maria Jeffs/EyeEm/Getty Images, 78; © Mary Doherty Photography/Getty Images, 26; © Mike Dunning/Getty Images, 37 bottom; © Mint Images — Norah Levine/Getty Images, 29; © Nancy Honey/Getty Images, 54; © northlightimages/Getty Images, 96; © OJO Images Ltd./Alamy Stock Photo, 56; © PIER/Getty Images, 15; © praetorianphoto/iStockphoto.com, 97; © Robert Houser/Getty Images, 10; © Rosalina Stevens/EyeEm/Getty Images, 77; © Rosl Roessner/BIA/Minden Pictures/Getty Images, 124; © Sandra Lundy Photography/Getty Images, 87; © Susanne Friedrich/Getty Images, 118; © Thom Lang/Getty Images, 120; © Tim Graham/Getty Images, 127; Vicki Stiefel, 47; © Walter B. McKenzie/Getty Images, 28; © wee2beasties/Getty Images, 142; © Zoltan Molnar/Alamy Stock Photo, 64